ALICIA'S DISCOVERIES
Stars

By Catarina Cunha
Illustrated by Megumi Arai
With Abbigail Elms

"This is for my dear Alicia.
Keep asking all the questions
because curiosity is the path of adventure
that leads to discoveries."
Catarina Cunha

Alicia's Discoveries

Stars

By Catarina Cunha
Illustrated by Megumi Arai
With guest Abbigail Elms

It is Friday evening, and Alicia is trying to find her favorite flashlight with Bigo's help. Alicia exclaims: "There it is, Bigo! It was under the stuffed animals."

They both run to Alicia's mom:
"We are ready to go to the park!" Tonight they are going to the observatory in the park to watch and study the stars with a giant telescope. Alicia's mom says: "Perfect, I have the blankets and the bottle with hot tea. Let's go!"

Alicia says in the car: "I'm so glad today the sky is clear! We will see so many stars!"

Bigo holds his head happily out of the window. Alicia's mom answers: "Yes, and the best part is that the astrophysicist Abbigail Elms is going to be there, too!"

Alicia checks the map at the park entrance: "Here is where we are. The way to the observatory is through the wildflower field." Alicia's mom answers: "Perfect! Let's go."

When they get close to the field, Alicia says: "I can already smell the scent of wildflowers! We must be close." Alicia walks ahead with Bigo, and Alicia exclaims in amazement: "Mom, look at all the fireflies! It looks like a magical fairy world! How do the fireflies light up at night?" Alicia's mom: "Yes, this is a stunning scene!

Fireflies light up because of a chemical reaction with luciferin inside their bodies.
Bioluminescence is the name for this kind of light-making. A firefly's light is "cold light" because it doesn't waste energy as heat. This is important because a firefly wouldn't be able to live if the part of its body that makes light got as hot as a light bulb."

TRANSPARENT EXOSKELETON
Light passes through the insect's see-through, hard outer covering.

REFLECTOR CELLS
Direct light outward

LIGHT CELLS
Where the chemical reaction occurs.

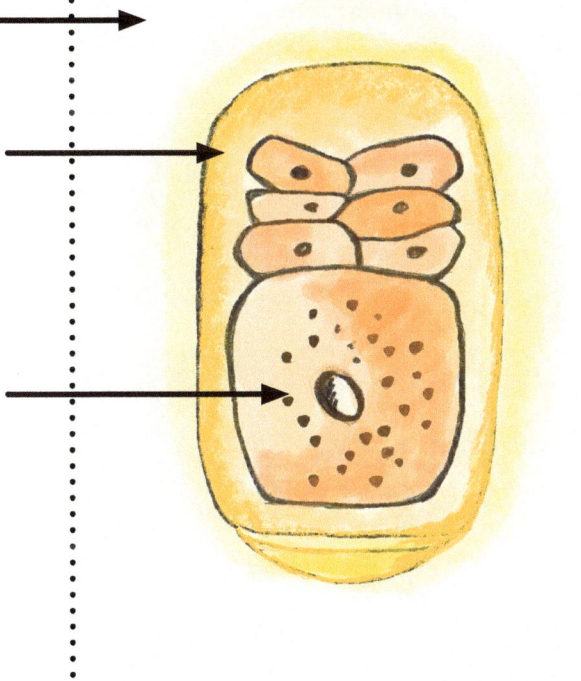

The observatory is in the middle of the wildflower field, and when they arrive, Abbigail greets them: "I'm so happy that you are here! I will be talking about my star research today!" Alicia says: "Thank you, Abbigail. I love stars!"

They all walk to the giant telescope, and Alicia gets a turn to look through it. While Abbigail explains: "Does anyone know why stargazing is like looking back in time?

The bright star Sirius is a distance of 8.6 light years away from us. That means the light hitting your eye tonight from Sirius has been traveling for 8.6 years. When you look at Sirius tonight, you see how it looked 8.6 years ago!"

The effect gets bigger and bigger as you look at things farther away. The Big Dipper's stars are between 60 and 125 light years away. When you look at Dubhe, the first star in the "bowl" of the Big Dipper, you see the light from before you were born." Alicia asks: "What are stars, and where do they come from?"

Abbigail answers: "That is a fascinating question, Alicia. The field of study to answer how a star changes over time is called "stellar evolution." Between the time they are made and the time they die, stars can change in many ways. Scientists like myself study stellar evolution by looking at many different stars at different times in their lives. This is because stars can give off light and heat for millions or billions of years.

A star goes through four stages in its life: a nebula, a main-sequence star, a red giant, and either a white dwarf, a neutron star, or a black hole.

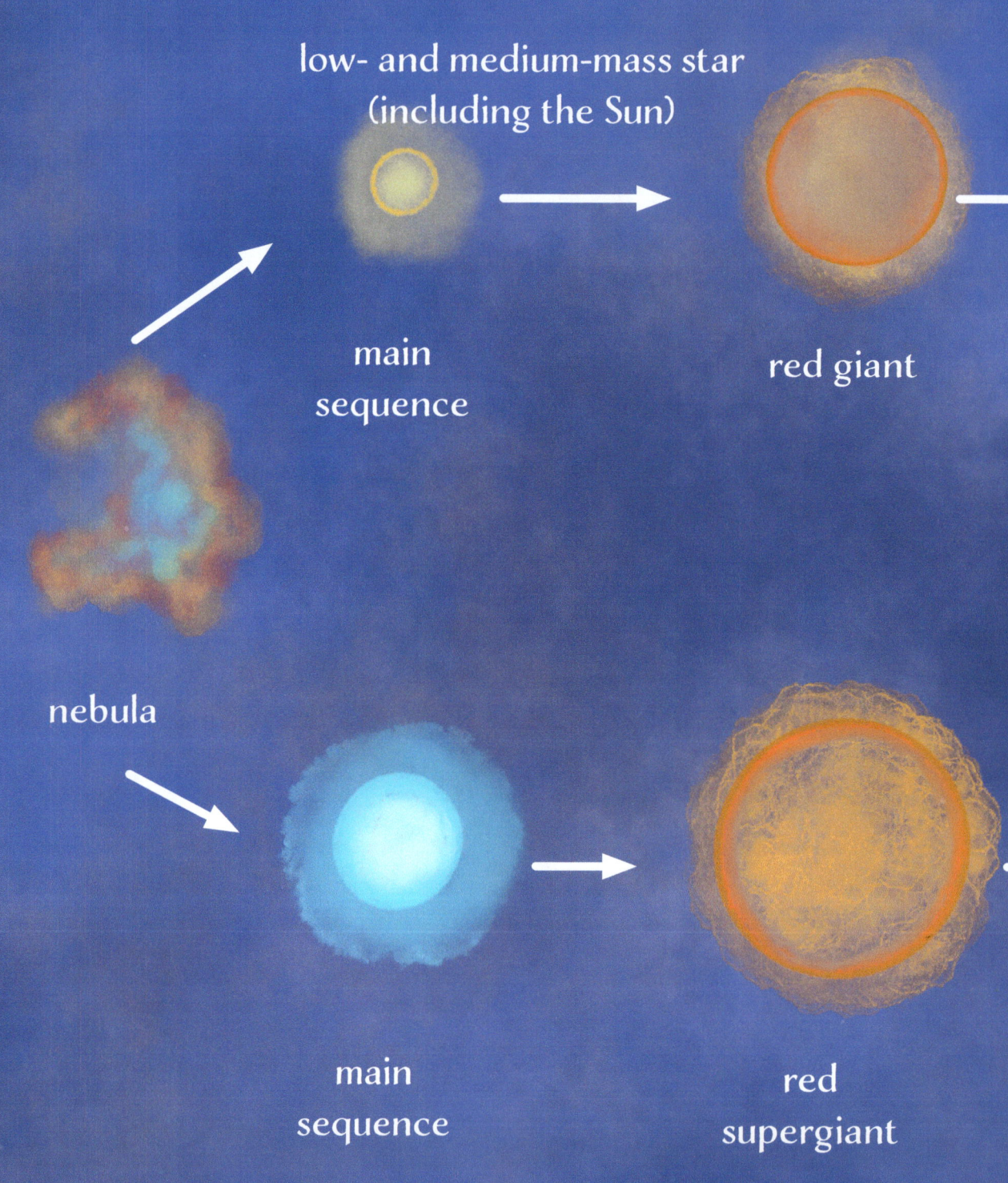

low- and medium-mass star (including the Sun)

main sequence

red giant

nebula

main sequence

red supergiant

I am very interested in the white dwarf stage of a star. White dwarfs can tell us about the formation and evolution of our entire galaxy, which helps us understand how the Earth was made.

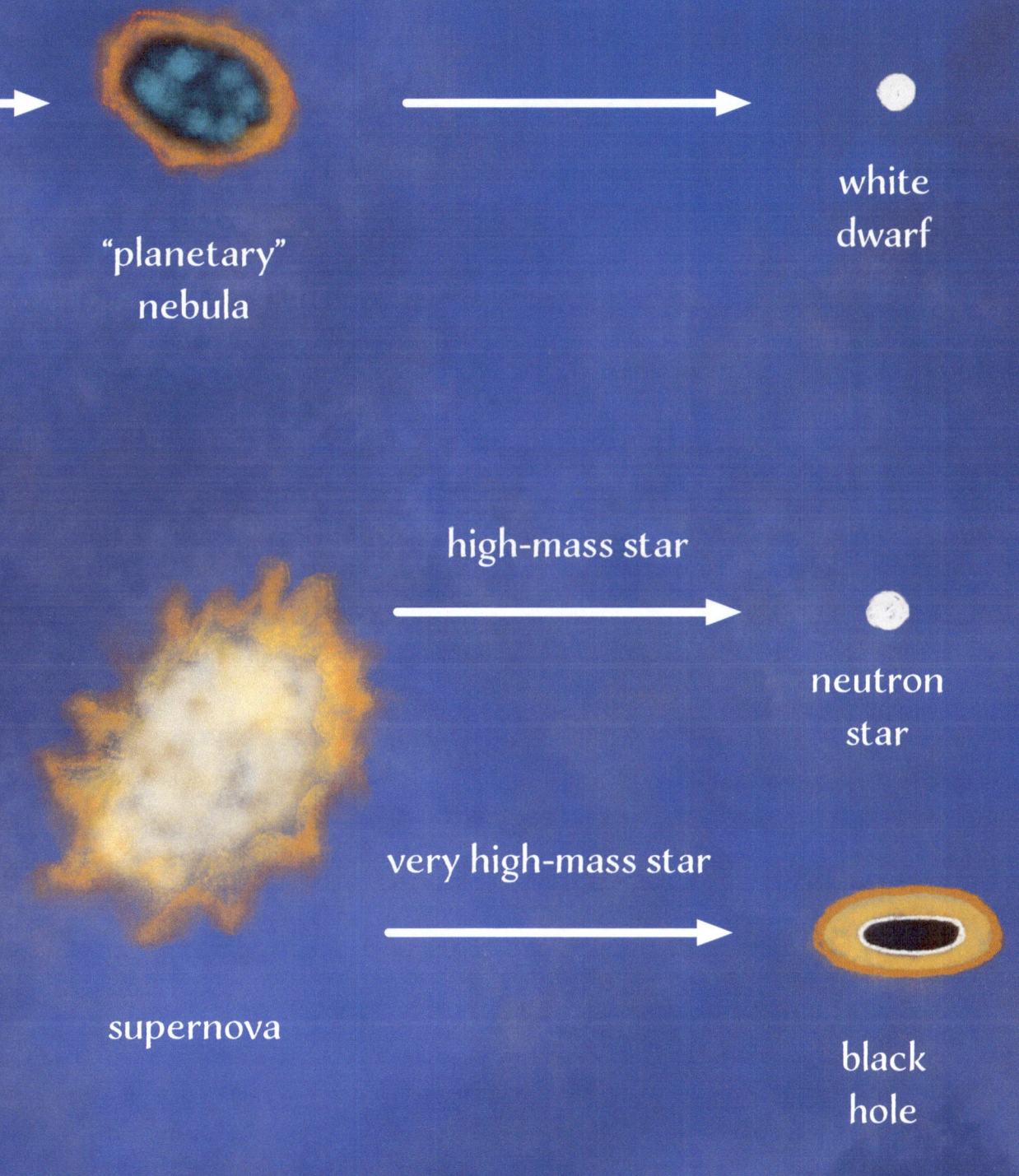

Over 97% of the Milky Way's stars will turn white dwarf stars.

When stars like our sun run out of fuel, they turn into a white dwarf.

"Near the end of a star's life, it will go through a red giant stage and then push its gas out into space. All that's left is the core of the star which is a young white dwarf. White dwarfs then grow old and cool over billions of years. I recently discovered the oldest polluted white dwarf in our galaxy."

Alicia asks: "How did it get polluted? Did it also have too many cars in cities?" Abbigail smiles:
"That's a great question. Polluted white dwarfs are white dwarf stars that have swallowed
orbiting planets and asteroids.

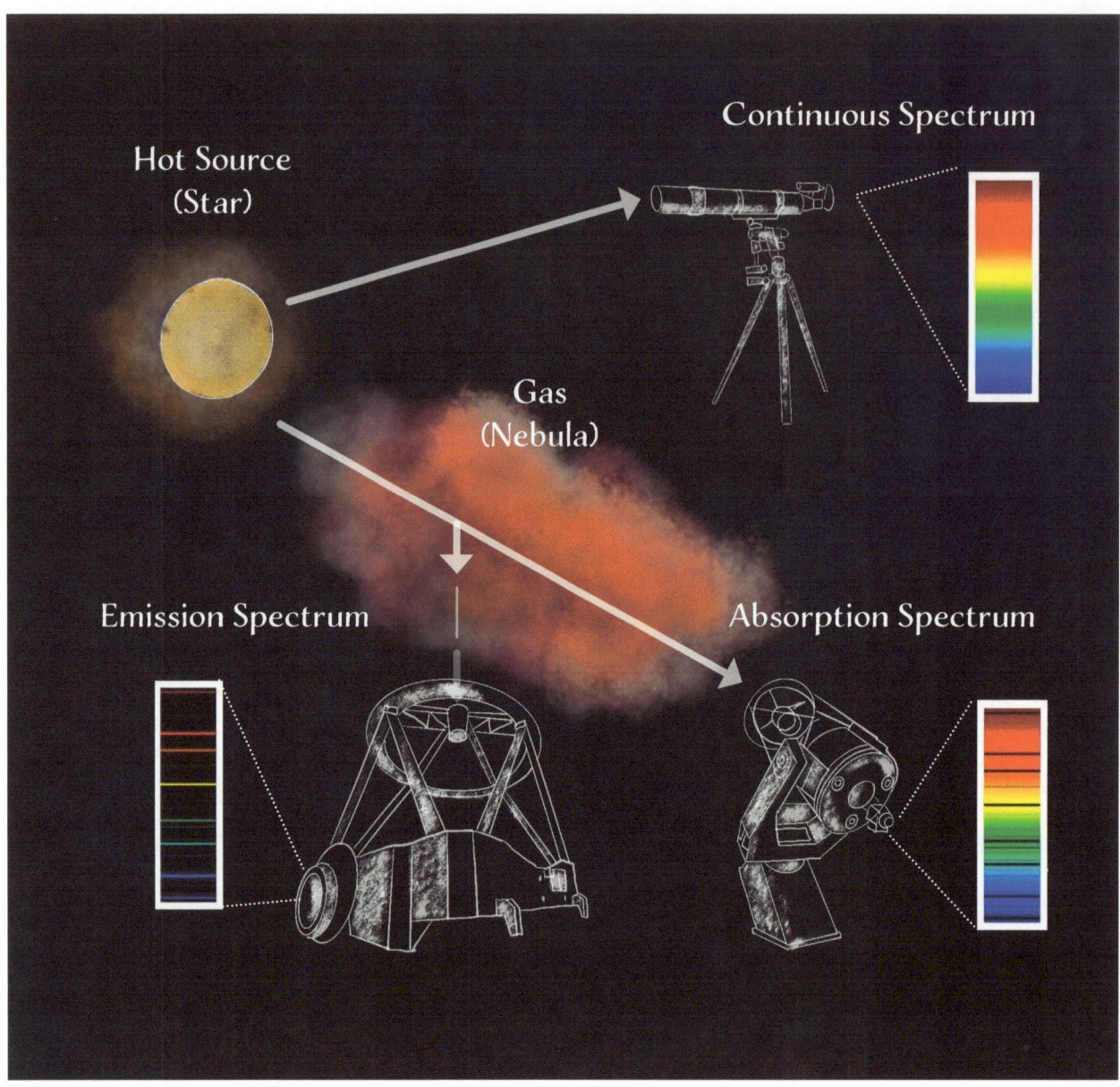

We used spectroscopic observation techniques that break up the light from a polluted white dwarf into its parts. I then used computer models to determine how much of each chemical was there to see what asteroids and planets were made of before they burned up in their atmospheres. This study is an essential piece of the puzzle for figuring out how our galaxy looked when it was young."

Alicia says: "Wow, that is amazing! I wish I could look at my favorite stars at home, but the city is too bright at night." Abbigail answers: "That is why we are making our favorite star constellations with the flashlights I asked you to bring." Alicia says: "You're the best, Abbigail!"

Materials:

What we need:

Round Flashlight

Thin cardboard

Glue

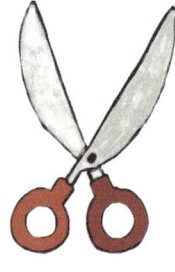
Scissors

Printed favorite constellation on paper (example):

A tool to pierce paper (a push pin, thumb tack, or nail)

Instructions:

1. Cut out the constellation in a circle with scissors.

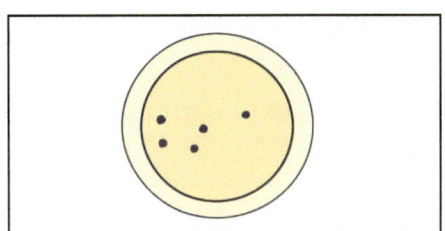

2. Stick the circles onto the cardboard to make them stronger.

3. Once the glue is dry, you can cut the circles with scissors.
Carefully make a hole in each star on the circle of the constellation. (An adult should use scissors or a push pin to do this step.)paper.

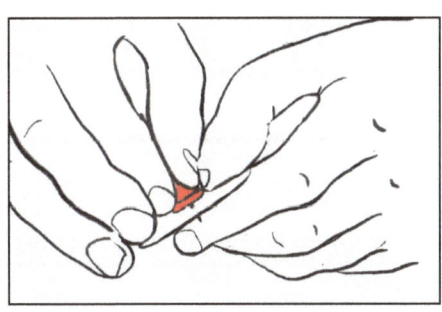

4. Make sure the picture faces the flashlight so that it shows up on the wall.

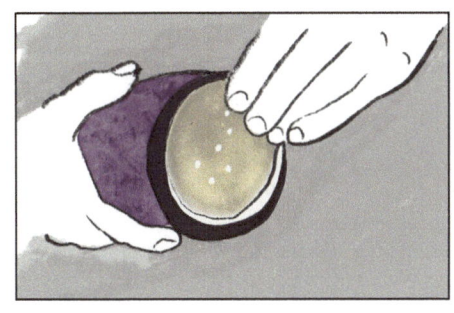

5. Shine the flashlight on a wall in a dark room to see the points of light that make up the constellation.

Back at home, Alicia projects her favorite star constellation onto the ceiling of her room and falls asleep, dreaming of traveling through the galaxy as the first kid astronaut.

I hope you all enjoy looking at the stars in your room!

Copyright © 2023 by Catarina Cunha
All rights reserved. This book or any portion thereof
may not be reproduced or used in any manner whatsoever
without the express written permission of the publisher
except for the use of brief quotations in a book review.

www.ingramcontent.com/pod-product-compliance
Lightning Source LLC
Chambersburg PA
CBHW041935240526

45473CB00034B/1710